# Health 98

# 根与芽
## Roots and Shoots

*Gunter Pauli*

[比] 冈特·鲍利　著

[哥伦] 凯瑟琳娜·巴赫　绘

何家振　译

上海远东出版社

# 丛书编委会

主　任：田成川

副主任：何家振　闫世东　林　玉

委　员：李原原　翟致信　靳增江　史国鹏　梁雅丽

　　　　任泽林　陈　卫　薛　梅　王　岂　郑循如

　　　　彭　勇　王梦雨

特别感谢以下热心人士对童书工作的支持：

匡志强　宋小华　解　东　厉　云　李　婧　庞英元

李　阳　刘　丹　冯家宝　熊彩虹　罗淑怡　旷　婉

杨　荣　刘学振　何圣霖　廖清州　谭燕宁　王　征

李　杰　韦小宏　欧　亮　陈强林　陈　果　寿颖慧

罗　佳　傅　俊　白永喆　戴　虹

# 目录

# Contents

一些细菌被困在牙齿的龋洞里，因为上面被金属盖盖住了，他们出不来。

一个细菌说："看看这个可怕的、黑乎乎的地方，我们被困在这里了！我们只不过是想吃那些粘在牙齿上的甜食而已。"

Some bacteria are trapped in a tooth cavity under a metal cap and cannot get out.

One says, "Look at this scary, dark place where we've ended up! And all we wanted to do was to eat the sweet food rests sticking to this tooth."

一些细菌被困在牙齿的龋洞里

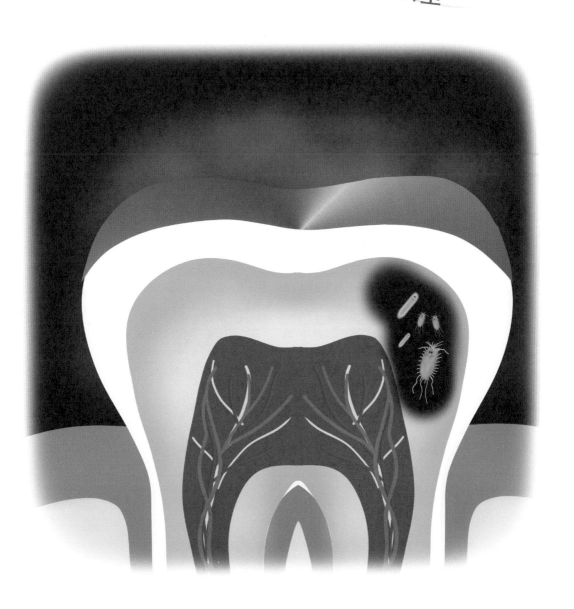

Some bacteria are trapped in a tooth cavity

这里除了骨头，什么都没有

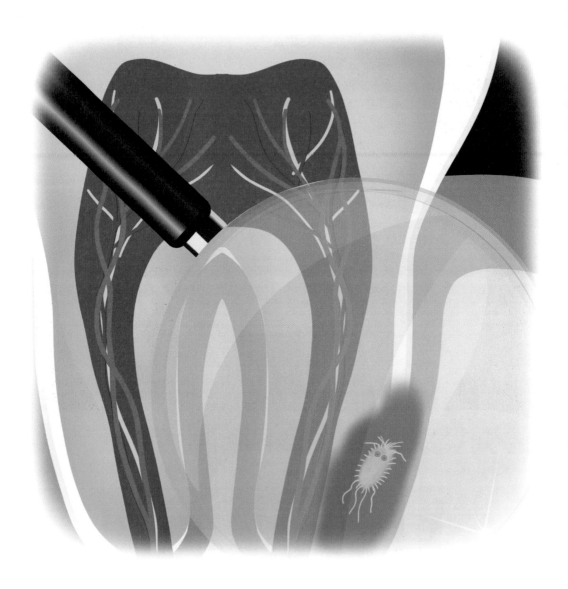

Nothing but bone around here

"现在我们被困在牙齿里这么深的地方。这里好像除了骨头什么都没有。"另一个细菌说。

"那一定是因为我们太贪吃了，想把能找到的东西都吃掉。现在可好，只剩下骨头和金属这些硬邦邦的东西了。"

"Now ow we are stuck deep inside this tooth. It feels like there's nothing but bone around here," says another.

"We e must have been too greedy trying to eat everything we could find. And now all that's left is this hard stuff – bone and metal."

"你总是起劲儿地挖，把所有龋洞里的糖渣挖得一点儿不剩！"他的朋友指责道，"我们一直挖呀挖，到处打洞，几乎所有牙齿都被我们打通了。"

"嗯，这些洞被称为龋洞。我们通常都会把洞打得这么深。你大概不会相信，能做到早餐前刷刷牙，把我们踢出来的人真不多。"

"You're the one that was so keen to get to the last bit of sugar out of every hole!" his friend accuses him. "We've been digging and digging through everything until there was hardly any tooth left."

"Well, these holes are called cavities. It's normal for us to tunnel our way through here. You won't believe how few people brush their teeth early in the morning to kick us out before they have breakfast."

刷牙的人很少能……

How few people brush their teeth ...

我喜欢这些吃剩的饭菜

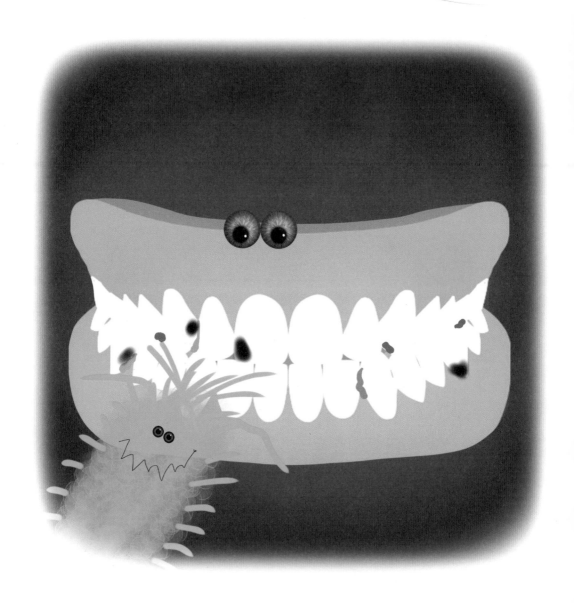

I love these left-overs

"那有什么不好呢？你不想吃他们吃剩的饭菜吗？"

"我喜欢这些吃剩的饭菜。不过，这里的细菌太多了，我们得快点儿吃。"

"So what's wrong with that? Don't you want to eat all their left-overs?"

"I love these left-overs. But as there are so many of us around here, we'd better get everything as quickly as we can."

"我知道这里的细菌很多。"他的朋友答道，"我听说，一个不注意口腔卫生的人嘴里的细菌数量比地球上的总人口还多。"

"哈哈，这就是不注意口腔卫生的人口臭严重的原因！因为我们在他的舌头、牙龈和牙齿上举办晚会呢。"

"I know that there are a lot of us," replies his friend. "I've been told that there are more bacteria in the mouth of a person who neglects his teeth than there are people on earth."

"Ha-ha, that must be why this guy has such bad breath! It's because we're having a party on his tongue, gums, and teeth."

在他的舌头、牙龈和牙齿上举办晚会

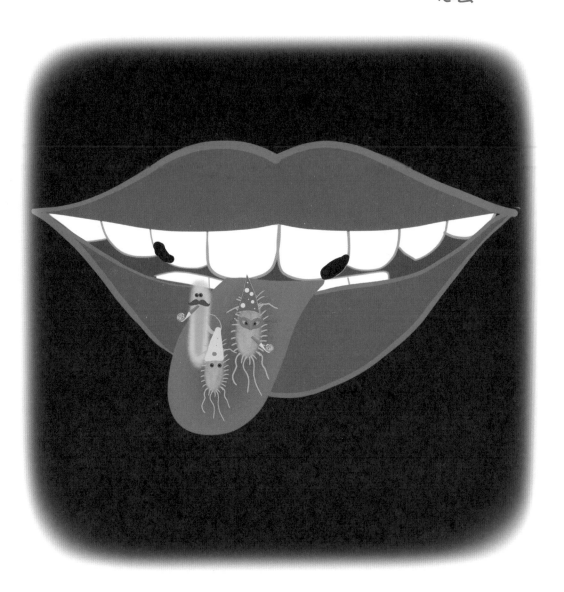

Having a party on his tongue, gums and teeth

坐在一个牙根管里

Sitting in a root canal

"但是自从我们被逼到这个洞底，快乐的时光好像彻底结束了。"他的朋友回应道。

"我觉得我们正坐在一个牙根管里，以前这里可是牙神经待的地方。或许这是一个移植管，我分辨不出来。我能肯定的是这里根本没有糖。"

"But party time seems to be over for us since we've been pushed deep down this hole," replies his friend.

"I think we're sitting in a root canal, where the nerve used to be. Or maybe it is an implant. I cannot tell. What I do know is that there's certainly no sugar down here."

"我想，牙主人的牙被我们吃掉了这么多，他一定疼得厉害，所以牙医不得不把他的牙神经抽出来。"

"我们肯定就是这样被逼到这里来的。"

"牙医确实想用他的刺激性化学药品杀死我们，但是你知道的，人们越想杀死我们，我们中活下来的细菌就会变得越强。"

"You know, I think we ate so much of our host's tooth away that it started hurting so badly that the dentist had to take the nerve out."

"And that must be how we got pushed in."

"The dentist did try to kill us with his harsh chemicals, but as you know, the more people try to kill us bacteria, the stronger those that survive will become."

牙医想用化学药品杀死我们

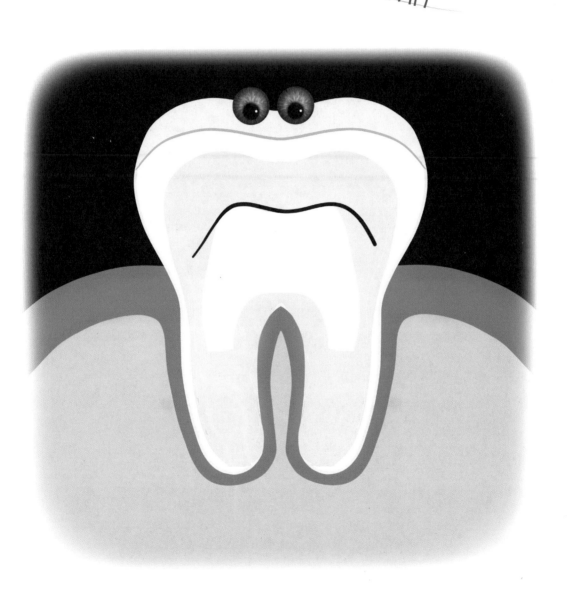

Dentist tried to kill us with chemicals

应该改变我们的饮食习惯

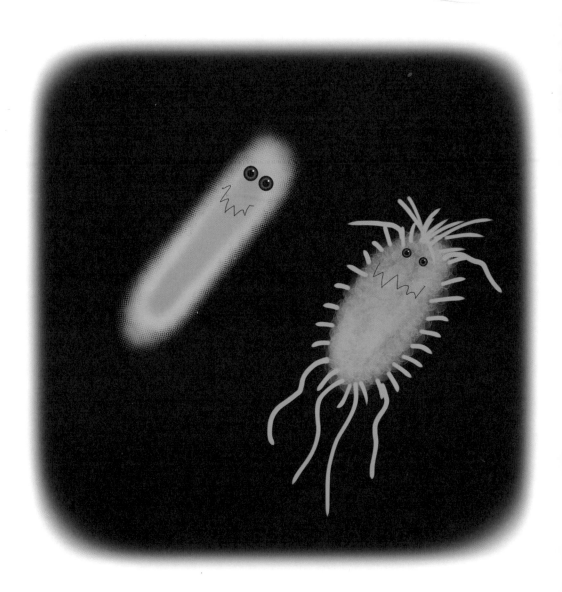

Better change our diet

"那么，你建议我们现在做点什么呢？"

"我想，我们应该改变我们的饮食习惯。"

"有什么用呢？在这里，我们无路可逃。上方的金属盖挡住了我们的去路，而且我们也没法穿过牙釉质或者牙根。"

"So what do you suggest we do now?"
"I think that we'd better change our diet."
"To what? We have no choice down here. And this metal cap above us is blocking the way out. And we cannot get out through the enamel or the roots either."

"因此我想，我们唯一的选择就是留在这里，吃掉牙齿的骨质部分，然后从牙齿的另一面逃出去。"

"那好，现在就开始享受我们的盛宴吧!"

……这仅仅是开始! ……

"Then I suppose the only option we have left is to eat away at the bone and hope we can one day get out on the other side."

"And start feasting again!"

... AND IT HAS ONLY JUST BEGUN!...

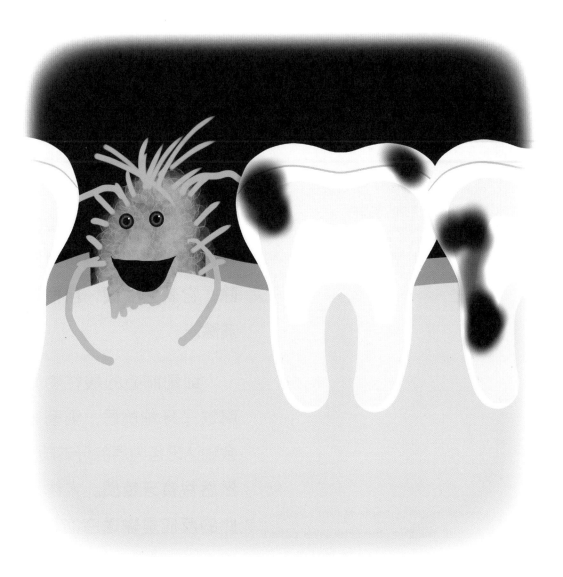

……这仅仅是开始！……

… AND IT HAS ONLY JUST BEGUN! …

When simple sugars and carbohydrates on the teeth are not cleaned off, bacteria eat them and produce acids. When many bacteria feed and multiply, they form plaque. If this sticky film is not removed when it is soft, it will get hard and be tough to get rid of.

如果单糖或碳水化合物残留在牙齿上，细菌就会吃它们，并产生酸性物质，当很多细菌都来吃并开始繁殖时，牙菌斑就形成了。如果人们没有趁黏性菌膜还软的时候将其清除，它就会变硬，且很难清除。

细菌制造的酸性物质腐蚀了牙釉质后，细菌就会到达牙齿内层的牙本质，然后吞食牙髓质。人体对此的反应是输送白血球，这可能会导致牙周脓肿。

The acids made by bacteria remove the tooth's enamel, reaching the inner layer called dentin. When bacteria reach beyond this, they devour the inner tooth pulp. The body responds to this by sending white blood cells and this may result in a tooth abscess.

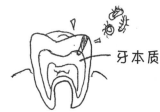

牙本质

During a root canal treatment, the infected pulp is removed, along with the nerve. The inside is cleaned and disinfected, and then filled with a filling and sealed with a crown.

在牙根管的治疗过程中，被感染的牙髓质与牙神经一起被移除。医生在牙根管内部进行清洁并消毒，然后填上填充物，并用金属盖封上。

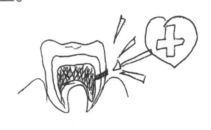

There are an estimated 20 billion bacteria (from 500 to 1,000 different families of bacteria) in our mouths. Some reproduce every five hours. Most of the bacteria are beneficial and aid in preventing disease.

据估计，我们的口腔里住着 200 亿个细菌（它们来自 500 到 1 000 个不同的细菌家族）。有些细菌每 5 小时繁殖一次。大多数细菌是有益的，有助于预防疾病。

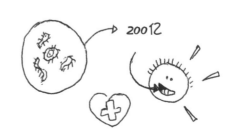

口腔里的有益细菌是我们免疫系统的第一道防线。如果没有它们，大量的细菌将通过空气和唾液的传播进入我们体内，人体将无法承受。有益菌还能抑制真菌的增长。

Oral bacteria are the first line of defence of our immune system. Without these good bacteria, our body would be overloaded with airborne and saliva-transferred germs. These good bacteria also control the growth of fungi.

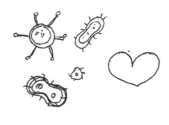

细菌产生的酸性物质会导致微创伤，唾液里的钙离子能修复这些创伤。唾液里的蛋白质能抑制细菌，防止其沾在口腔内部，然后唾液会将其冲走并吞掉。

Saliva in the mouth contains the calcium ions needed to repair the minimal damage caused by bacterial acids. The protein in saliva binds bacteria, preventing it from sticking to any surface, and it is then washed away by the saliva and swallowed.

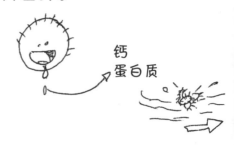

钙
蛋白质

Brushing and flossing teeth prevent, or at least reduce, infection and disease. However, reducing the intake of sugars and carbohydrates in our diet has a greater influence on oral health.

刷牙或者用牙线洁牙，能够防止或减少感染和疾病。但是，减少糖和碳水化合物的摄入更为重要。

It is best to brush your teeth in the morning before breakfast, as the greatest amount of plaque is present in our mouths when we wake up. It is also easier to remove it then.

每天早餐前最好刷刷牙，因为在刚醒的时候牙菌斑最多，而且最容易清除。

你认为你能清除口腔里所有的细菌吗?

Do you think you can ever get rid of all the bacteria in your mouth?

如果你深陷一个大坑中，而且被封在里面，你是否准备为了生存而改变你的饮食习惯?

If you were trapped in a big hole that is sealed off, would you be prepared to change your diet to survive?

当有很多食物任你享用时，你是否比平常吃得更多呢?

When there is a lot of food available, would you eat more than you usually do?

你认为是细菌太贪吃，理应被困在牙齿深处，还是牙主人吃的食物有问题呢?

Do you think the bacteria were greedy and therefore to blame for their predicament or did the person eat the wrong food?

Do you have your microscope ready? Ask a parent or teacher to assist you in scraping some plaque off your teeth with a metal scraper. Put it on a glass slide and have a look through the microscope at the live bacteria moving around. Is the sight of these micro-organisms, fresh off your teeth, motivating you to take better oral care?

准备好你的显微镜了吗？请一位家长或老师用金属刮刀帮你刮掉牙齿上的牙菌斑。将牙菌斑放在玻璃片上，通过显微镜观察活动的细菌。观察这些刚从你牙齿上清除下来的微生物，会促使你做好口腔护理吗？

# 学科知识

## Academic Knowledge

| | |
|---|---|
| 生物学 | 口腔里最常见的两种有害菌是变异链球菌和牙龈卟啉单胞菌；细菌通过二分裂法生长；因为没有核，细菌会快速地发生突变；不好的气味来自舌头上的小裂隙，导致口臭的细菌在那里进行繁殖；多数产生口臭的细菌是厌氧菌，它们在人们闭上嘴睡觉的时候生长得最快。 |
| 化 学 | 溶菌酶是人唾液中的抗菌物质，它能杀死细菌，或者至少可以抑制其活性；唾液中含有磷酸盐和钙离子，能够帮助修复由细菌酸引起的牙齿微损伤；我们唾液中的维生素K可以防止酸性物质的形成，人们在发酵食物（如德国泡菜）中发现了维生素K，蛋黄和食草动物的制品中也有维生素K；唾液酸性越强，龋齿就越多。 |
| 物 理 | 人在一天中通过咀嚼食物会产生15 000次电脉冲，可使身体保持活力，因为通过神经网络，每颗牙齿都与特定器官相连。 |
| 工程学 | 牙科医生不断探索贵金属合金和新瓷材料的结合，以提高填充材料和移植物的强度和弹力。 |
| 经济学 | 口腔卫生不良的代价很大，它会增加患心脏病、中风、痴呆、呼吸系统疾病、糖尿病和癌症的风险。 |
| 伦理学 | 贪婪对你和他人的生活会造成很大的影响；人类倾向于摧毁他们不理解的东西，结果导致有益菌被杀死。 |
| 历 史 | 荷兰科学家安东尼·范·列文虎克是第一位用显微镜观察细菌的人，当时他在研究牙斑；3 500年前人类就开始使用牙刷了；牙刷的工业化生产始于15世纪的中国；在19世纪，家庭制作的牙膏中含有木炭粉。 |
| 地 理 | 波兰儿童12岁之前平均掉4颗恒牙，而相同年龄的德国和英国儿童掉牙不超过一颗。 |
| 数 学 | 口腔中的细菌在很短的时间内以指数方式增长。 |
| 生活方式 | 除了有规律地刷牙，改变饮食也是对牙齿很好的保护：少吃单糖和碳水化合物；牙刷已经成为一次性物品；牙刷毛是由尼龙制成的，降解非常慢，半衰期长达几百年。 |
| 社会学 | 总体而言，中等收入家庭成员的牙齿健康最糟糕，而贫困家庭和富裕家庭稍好一些。 |
| 心理学 | 蛀牙的发展与消极情绪、压力之间的关系。 |
| 系统论 | 少吃精加工的糖和精白面粉可以降低蛀牙的概率；免疫系统的第一道防线是口腔里的有益菌和唾液。 |

## 情感智慧
### Emotional Intelligence

细　菌

细菌肆无忌惮，充满自信。他们认识到其生存环境已经发生了根本变化。他们质疑自身的行为，并且想知道自己忘乎所以地吃掉所有能吃的东西是否太贪婪了。细菌很清楚人们知道如何摆脱他们，但是多数人懒得去做。细菌明白发生了什么，并决心生存下去，即使这意味着要改变饮食习惯。他们怀抱希望，相信很快就能恢复原有的习惯，继续他们数百万年来的生活方式。

## 艺术
### The Arts

如何处理旧牙刷？很多旧牙刷进了垃圾填埋场。让我们收集旧牙刷，并将它们变成油画刷！我们用牙刷画幅画吧。就拿旧牙刷作为你的画笔，蘸上油画颜料，在厚纸板或者帆布上画吧。这些旧牙刷可能没剩多少毛可以刷牙了，但是用来当油画刷已经足够了。这给你提供了一个利用旧物表达自我的有趣方式。

## 思维拓展
### Systems: Making the Connections

　　口腔是人体与外部世界之间的第一道主要屏障，因此是人类免疫系统的第一道防线。如果没有口腔提供的独特保护，数十亿通过空气传播的细菌和真菌将进入人体。作为防御机制，免疫系统与其说是一个杀菌机器，不如说是一个不断寻求平衡和排除的过程。摄入过多单糖和碳水化合物的现代饮食习惯，增加了免疫系统的负担。唾液中的钙离子自然修复系统，无法负担过量"细菌食物"的负荷。过多的不健康食品引来了对我们的健康存在潜在威胁的细菌。为了保持最好的口腔卫生状态，必须经常刷牙并用牙线清洁牙齿。理想的解决方案是大量减少糖和碳水化合物的摄入。除了作为阻挡细菌入侵的屏障，口腔和牙齿还经过神经系统与人体其他器官相连。咀嚼本身就能激活和刺激身体，掉一颗牙齿就会失去部分刺激功能。金属的引入，从人造牙冠、填充物到人工植牙，都会阻碍人体的能量流动。而这种能量流动对保持身体和免疫系统处于激活状态并持续发挥功能是非常必要的。如果人们长期不健康饮食，细菌也会改变它们的饮食，从而导致失衡，摧毁健康系统的根基。如果链球菌过多、真菌失衡、骨质被细菌吃掉，牙齿就会失去刺激身体其他部位的能力。重金属从口腔渗透体内，使身体变得不堪重负，甚至中毒，需要专业手段才能解毒。金属等外来物质应该从口腔里清除，否则它会持续释放微量金属，增加免疫系统的负担，甚至导致免疫系统崩溃。从改变饮食习惯开始，清除各种来源的重金属，保持口腔卫生，恢复牙齿对各器官的刺激功能，这才是通往健康和幸福真正的系统方法，远比刷牙和用牙线洁牙更有意义。

## 动手能力
### Capacity to Implement

　　组织一次刷牙音乐会。世界上第一次刷牙音乐会是在尼日利亚举办的，30万名儿童同时开始刷牙。你可以在学校或家里举行一次这样的活动。确保每个人都有一把好牙刷，一把很耐用的牙刷。放点音乐，跟着节奏刷掉牙菌斑。当所有人做好刷牙的准备后，喊："预备，开始！"仔细地将牙刷上下（垂直）移动以确保不会损害牙龈。只有当牙刷放在你牙齿的表面上时，才能左右（水平）移动。你可以用牙线洁牙来结束你的刷牙音乐会。

## 故事灵感来自
### This Fable Is Inspired by

# 托马斯·拉乌医生
## Dr Thomas Rau

托马斯·拉乌是一位医师，他毕业于伯尔尼大学。在一家康复中心工作时，他发现尽管他按照学校所教的知识尽心地给病人治疗，但是病人的病情并没有好转。于是拉乌医生转而成为印度草药按摩和中医的终身学习者。从那以后，他成为整体论医学的拥护者，并阐述了一个理论：健康的恢复建立在中西医结合的基础上，包括解毒、营养、消化和提高免疫力。他遵循生物医学的原则，把口腔卫生和无金属牙科治疗融入他的医疗实践。拉乌医生现在是瑞士帕拉塞尔苏斯诊所的医疗总监。

**图书在版编目（CIP）数据**

冈特生态童书.第三辑修订版:全36册:汉英对照 /
(比)冈特·鲍利著;(哥伦)凯瑟琳娜·巴赫绘;
何家振等译.—上海:上海远东出版社,2022
书名原文:Gunter's Fables
ISBN 978-7-5476-1850-9

Ⅰ.①冈… Ⅱ.①冈… ②凯… ③何… Ⅲ.①生态环
境–环境保护–儿童读物—汉、英 Ⅳ.①X171.1–49

中国版本图书馆CIP数据核字(2022)第163904号
著作权合同登记号图字09-2022-0637号


| 策　　划 | 张　蓉 |
| 责任编辑 | 程云琦 |
| 封面设计 | 魏　来　李　廉 |


冈特生态童书
**根与芽**
[比]冈特·鲍利　著
[哥伦]凯瑟琳娜·巴赫　绘
何家振　译

记得要和身边的小朋友分享环保知识哦！
八喜冰淇淋祝你成为环保小使者！